Once there were three billy goats Gruff.

The goats strolled along a road. They wanted to find some grass to eat.

1

The old goat saw green
grass. "Let's go there," he told
the others. "We'll see who can
find the most to eat!"

The little goat started to
cross the plank.

Under the plank was
a wild troll. "Hold on!" the
troll croaked. "I'm going to
gobble you up!"

"I'm too little to eat," said
the goat. "Wait for my brother.
He is quite big."

"Go then!" croaked the troll.

4

The next goat started
over the plank. "I'll eat **you**!"
croaked the troll.

"Wait for my brother," the
goat told him. "He will make a
better meal."

Then the old goat started
over the plank.

"Hold on!" croaked the troll.
"I'm going to gobble you up!"

"I'm big and bold," boasted
the goat. "You won't eat me!"
The old goat kicked the
troll. The troll landed in the
cold stream.

The three goats roamed
the hills. "This grass is the
best kind!" they said. "We are
so fat that we roll!"

They stayed there and were
most happy.

The End